GSAS
PUBLICATIONS SERIES

DISTRICTS: GSAS ASSESSMENT

v2.0 - 2013

Dr. Yousef Al Horr
Founding Chairman

A MESSAGE FROM

DR. YOUSEF MOHAMED AL HORR,
FOUNDING CHAIRMAN

GORD has come a long way since pioneering the Global Sustainability Assessment System (GSAS), formerly known as (QSAS), the Middle East's first integrated and performance-based green building assessment rating system in 2009.

Our mission to encourage the development and implementation of sustainability principles and imperatives stems from the sustainable goals outlined in His Highness, The Emir Sheikh Hamad bin Khalifa Al-Thani's Qatar National Vision 2030, which aims to achieve sustainable economic development and environmental leadership.

GSAS draws from top-tier global sustainability systems and adds new facets and dimensions to the current practices in assessing the sustainability of the built environment. Modelled on best practices from the most established global rating schemes including, but not limited to, BREEAM (United Kingdom), LEED (United States), GREEN GLOBES (Canada), CEPAS (Hong Kong), CASBEE (Japan), and the International SBTOOL, GSAS has grown into a pan-regional system offering a comprehensive framework, and equally flexible to incorporate the specific needs of the local context of different regions. In Qatar, GSAS is currently the only rating scheme to be acknowledged by Qatar Construction Specifications (QCS 2010).

Primary goals of GSAS include creating a better living environment, minimizing resource consumption and reducing environmental degradation due to the fast pace of urbanization taking place in this era. Such objectives, coupled with the increasing evidence of climate change effects on a global level, have contributed strongly to the unprecedented pace of adaptation to sustainability practices not only in the developed countries, but also in developing countries at a pace that is unexpected.

GSAS Version 2.0 has become the most comprehensive system, to date, that addresses the built environment from a macro level to a micro level targeting a wide range of building typologies. The new system will have design assessments for all typologies integrated into one comprehensive manual. The manual provides recommendations and guidelines for the effective implementation of the sustainability goals of each criterion. As more research is carried out on the rating system, the manuals will be further developed to keep users informed on updates within the constantly evolving GSAS rating systems.

I would like to acknowledge the efforts and contributions from the State of Qatar, all our members, and international partners-especially the TC Chan Center for Building Simulation and Energy Studies at the University of Pennsylvania and the Center's associated consultants who helped establish the system and take it into new dimensions. Last but not least, the continuous support from Qatari Diar Real estate Investment Company is highly appreciated, and without its support, GSAS would not be able to achieve what it has achieved in such a short time.

DISTRICTS ASSESSMENT

TABLE OF CONTENTS

ACKNOWLEDGMENTS

GLOBAL SUSTAINABILITY ASSESSMENT SYSTEM (GSAS)
ACKNOWLEDGEMENT PREFACE

THIS PROJECT WAS INITIATED, COMMISSIONED AND LED BY

Dr. Yousef M Alhorr,
Founder and Chairman,
Gulf Organisation for Research and Development

PRINCIPAL PROJECT DEVELOPER AND DIRECTOR

Dr. Ali Malkawi
Professor of Architecture and Chairman of the Graduate Group, University of Pennsylvania,
Founder and Director, T.C. Chan Center for Building Simulation and Energy Studies

SPECIAL ACKNOWLEDGMENT

HE. Mr. Ghanim Bin Saad Al Saad
Chairman and Managing Director - Barwa

Eng. Mohammed Alhedfa,
GCEO, Qatari Diar Realestate Investment Company - State of Qatar

Dr. Mohammed Saif Alkuwari,
Under Secretary of Ministry Of Environment - State of Qatar

TECHNICAL LEAD AND DEVELOPER

Dr. Godfried Augenbroe,
Chair of Building Technology, Doctoral Program,
Professor, College of Architecture

DIRECTOR RESEARCH & DEVELOPMENT

Dr. Esam Elsarrag,
Gulf Organisation For Research & Development

ACKNOWLEDGMENTS

GULF ORGANISATION FOR RESEARCH AND DEVELOPMENT - QATAR

Team Leads

Hassan Satti   Ali Diyab Elsheikh

Technical Team

Abdulrahim Alsayed	Ronaldo Dalistan	Lakshmi Suryan
Omeima Khidir	Hassan El Aref	Mutasim Salim
Alek Zivkovic		

Logistic and Support Lead

Salah Al Ayoubi

Logistic and Support Team

Ibrahim Aburhaiem	Murad Ali Naz	Khalid Radi
Mohammed Imran	Mohammed Dad	Mohammed Elsheikh
Melody Manaman	Gloria Pineda	

TC CHAN CENTER – UNIVERSITY OF PENNSYLVANIA - USA

Team Leads

Chau Nguyen, Project Manager   Yun Kyu Yi, Assistant Project Manager

Research and Development

Yasmin Bhombal	Bin Yan	Khaled Tarabieh
Sean Williams	Aroussiak Gabrielian	Lily Trinh Ciammaichella
Joseph Hoepp	Charles Nawoj	Alex Muller
Sarah Savage	Kristen Sterner	Alexander Waegel
Kristen Albee	Niketa Laheri	Jeremy Krotz
Rebecca Lederer	Ethan Leatherbarrow	Noelle Tay
Erin Lauer	Yoon Soo Lee	

Web Development

Brandon Krakowsky	Marcus Pierce	Sunil Kamat
Sibasish Acharya	Ruchir Jha	

ACKNOWLEDGMENTS

AFFILIATED RESEARCH INSTITUTIONS
GEORGIA INSTITUTE OF TECHNOLOGY - USA

Research and Development Team

Sang Hoon Lee   Yeonsook Heo   Fei Zhao
Reen Foley

QATARI GOVERNMENT AND SEMI-GOVERNMENT SECTOR

Barwa Real Estate Company (BARWA)
Lusail Real Estate Development Company (LUSAIL)
Ministry of Endowment and Islamic Affairs (AWQAF)
Ministry of Environment (MOE)
Ministry of Interior – Internal Security Forces (ISF)
Ministry of Municipal Affairs and Urban Planning (MMUP)
Private Engineering Office – Amiri Diwan (PEO)
Public Works Authority (ASHGHAL)
Qatari Diar Real Estate Investment Company (QD)
Qatar General Electricity and Water (KAHRAMA)
Qatar Museums Authority (QMA)
Qatar Olympic Committee (QOC)
Qatar Petroleum (QP)
Qatar Science and Technology Park – Qatar Foundation (QSTP)
Qatar University (QU)

QATARI PRIVATE SECTOR

Arab Engineering Bureau (AEB)   HOARE LEA Qatar (HLQ)
KEO International Consultants   Energy City Qatar (Energy City)

REGIONAL PROFESSIONAL ORGANISATIONS

State of Kuwait – Green Building Committee – National Codes Committee
Kingdom of Jordan – Jordanian Engineers Association
Republic of Sudan - University of Khartoum

ACKNOWLEDGMENTS

INTERNATIONAL EXPERT REVIEWERS AND CONSULTANTS

- **Dick Van Dijk, PhD [Netherlands]**

  Member of ISO TC163 Energy Standardization Committee, TNO, Institute of Applied Physics.

- **Frank Matero, PhD [US]**

  Professor of Architecture and Historic Preservation, University of Pennsylvania.

- **Greg Foliente, PhD [Australia]**

  Principal Research Scientist, CSIRO (Commonwealth Scientific and Industrial Research Organisation) Sustainable Ecosystems.

- **John Hogan, PE, AIA [US]**

  City of Seattle Department of Planning and Development, Member of ASHRAE.

- **Laurie Olin, RLA, ALSA [US]**

  Partner, OLIN Studio.

- **Mark Standen [UK]**

  Building Research Establishment Environmental Assessment Method (BREEAM) Technical work.

- **Matthew Bacon, PhD, RIBA, FRSA [UK]**

  Professor, University Salford - Faculty Built Environment and Business Informatics; Chief Executive, Conclude Consultancy Limited; and Partner, Eleven Informatics LLP.

- **Matt Dolf [Canada]**

  Assistant Director, AISTS (International Academy of Sports Science and Technology).

- **Matthew Janssen [Australia]**

  Director of Construction and Infrastructure and Environmental Management Services Business Units (KMH Environmental); formerly the Sustainability Program Manager for Skanska.

- **Muscoe Martin, AIA [US]**

  Director, Sustainable Buildings Industries Council (SBIC), USGBC board member.

ACKNOWLEDGMENTS

- **Nils Larsson [Canada]**

  Executive Director of the International Initiative for a Sustainable Built Environment (iiSBE).

- **Raymond Cole, PhD [Canada]**

  Director, School of Architecture and Landscape Architecture, University of British Columbia.

- **Skip Graffam, PhD, RLA, ASLA [US]**

  Partner, Director of Research, OLIN Studio.

- **Sue Riddlestone [UK]**

  Executive Director & Co-Founder of BioRegional, Co-Director of One Planet and M.D. of BioRegional MiniMills Ltd.

PREFACE

The primary objective of Global Sustainability Assessment System (GSAS) is to create a sustainable built environment that minimizes ecological impact while addressing the specific regional needs.

The GSAS manuals and documents developed to date include the following:

- Construction: GSAS Guidelines v2.0
- Construction: GSAS Assessment v2.0
- Districts: GSAS Guidelines v2.0
- Districts: GSAS Assessment v2.0
- GSAS Energy Application v2.0
- GSAS Training Manual: Commercial & Residential – Part I v2.1
- GSAS Training Manual: Commercial & Residential – Part II v2.1
- GSAS Technical Guide v2.0
- Health Care: GSAS Design Guidelines v2.0
- Health Care: GSAS Design Assessment v2.0
- Parks: GSAS Guidelines v2.0
- Parks: GSAS Assessment v2.0
- Railways: GSAS Design Guidelines v2.0
- Railways: GSAS Design Assessment v2.0
- RFP Preparation: GSAS All Typologies v2.0
- Sports: GSAS Design Guidelines v2.0
- Sports: GSAS Design Assessment v2.0
- Typologies: GSAS Design Guidelines v2.0
- Typologies: GSAS Design Assessment v2.0
- Typologies: GSAS Operations Guidelines v2.0
- Typologies: GSAS Operations Assessment v2.0

DISTRICTS ASSESSMENT

SCOPE

GSAS Districts is intended to evaluate the planning and design of urban development projects. Districts typically consist of various building typologies and include several components such as infrastructure networks, transportation networks and public or open spaces. GSAS Districts can be applied to any combination of buildings and any size of development.

Both new and existing districts can be assessed under GSAS Districts. New districts will be evaluated according to the design intent of their master plan and a provisional certificate will be issued if the project achieves at least a 1 Star rating. After the construction of the district, the project will undergo design verification and a final certificate will be issued based on the results. Existing districts will be evaluated based on the actual built environment, including any changes made through revitalization efforts.

Districts are typically composed of several different building typologies, many of which are evaluated through the GSAS Design rating systems. In order to evaluate the sustainability of the development as a whole, the assessment of certain criteria in GSAS Districts depends upon the scores of criteria in GSAS Design. These GSAS Districts criteria include GSAS Rated Typologies [S.12], Water Consumption [W.1] and the four criteria under Energy [E.2 – E.5]. The project will identify target scores for new districts and achieved scores for existing districts for the building typologies that can be assessed under GSAS Design.

Certain one-off facilities or buildings may consume significant resources but are necessary for the efficient operation of the district. Because these facilities help the project achieve its sustainability goals, they may be omitted from Energy [E.2 - E.5] and Water Consumption [W.1] assessments. These buildings include, but are not limited to, district cooling plants, solid waste recycling centers and wastewater treatment facilities.

All projects must complete the assessment process for each criterion that is applicable to their particular typology. Some exceptions may apply based on the unique conditions of the project, and such exceptions will be determined, on a case-by-case basis, by the Certification Authority based on requests or submittals from the project. Examples of these exceptions include, but are not limited to, the following:

- For a criterion where the measurement does not apply, the project will automatically earn a baseline score of 0 without needing to complete the measurement process such as performing simulations, completing the calculator, etc.
- For a criterion where, by definition of the measurement and scoring range, the project should logically earn a score of 3, then the project is exempt from the measurement process.

In most cases, the project should complete the measurement process as defined by the criterion. Exemptions will not be given based on pre-existing or inevitable conditions. For example, if a project selects a site with high ecological value and inevitably must degrade the site in order for any development to take place, then the project will receive a low score for degrading the site.

In order to facilitate the evaluation of certain quantitative measurements, GSAS calculators were developed for many criteria to compute the project's performance and determine a final criterion score. Because the calculators are normative measures developed with available standards and technologies, new technologies may not yet be addressed in the GSAS Calculators. GSAS will continue to perform studies of new technologies as they become available in order to recalibrate the Calculators for future versions.

CRITERIA SUMMARY

The following chart summarizes the criteria within GSAS Districts, their associated weights and the goal of each category.

No	Category / Criteria	Weights	Goals
UC	**Urban Connectivity**	**10.00%**	The district shall control its effect on the urban environment with regard to existing infrastructure and amenities.
UC.1	Transportation Load	4.23%	
UC.2	Proximity to Existing Districts	2.52%	
UC.3	Acoustic Conditions	0.73%	
UC.4	Solid Waste Load	2.52%	
S	**Site**	**20.00%**	
S.1	Land Preservation	2.04%	
S.2	Water Body Preservation	2.71%	
S.3	Habitat Preservation	2.04%	
S.4	Vegetation	1.63%	
S.5	Walkability	1.81%	The district's site shall control the environmental impact of urban development.
S.6	Bikeability	1.81%	
S.7	Desertification	1.63%	
S.8	Parking Footprint	0.90%	
S.9	Mixed Use	1.09%	
S.10	Crime Prevention	0.81%	
S.11	Public Space	0.81%	
S.12	GSAS Rated Typologies	2.71%	
E	**Energy**	**18.00%**	
E.2	Energy Delivery Performance	4.71%	The district's depletion of fossil energy over its service life shall be controlled.
E.3	Fossil Fuel Conservation	3.29%	
E.4	CO_2 Emissions	4.12%	
E.5	NO_x, SO_x, & Particulate Matter	5.88%	
W	**Water**	**16.00%**	The district's impact on the overall water consumption and its associated burden on municipal supply and treatment systems shall be controlled.
W.1	Water Consumption	16.00%	

Table 1   Districts Rating System Scope, Part I

No	Category / Criteria	Weights	Goals
M	**Materials**	**8.00%**	The district's ecological impact shall be controlled with regard to factors associated with material extraction, processing, manufacturing, distribution, use/re-use and disposal for the development of the infrastructure and the design of buildings.
M.1	Regional Materials	2.18%	
M.2	Responsible Sourcing of Materials	2.55%	
M.3	Recycled Materials	1.82%	
M.4	Materials Reuse	1.45%	
M.5	Life Cycle Assessment (LCA)	N/A	
OE	**Oudoor Environment**	**7.00%**	The district's outdoor environmental quality shall be controlled with regard to factors such as thermal comfort and air quality, air movement and acoustics.
OE.1	Heat Island Effect	2.00%	
OE.2	Adverse Wind Conditions	1.50%	
OE.3	Air Flow	2.00%	
OE.4	Acoustic Quality	1.50%	
CE	**Cultural & Economic Value**	**13.00%**	The district shall enhance cultural values and boost national and local economies.
CE.1	Heritage & Cultural Identity	5.20%	
CE.2	Support of National Economy	4.68%	
CE.3	Housing Diversity	3.12%	
MO	**Management & Operations**	**8.00%**	The district's systems and infrastructure maintenance and operations plans shall be defined.
MO.1	Construction Plan	1.52%	
MO.2	Management Plan	1.52%	
MO.3	Wastewater Management Plan	1.52%	
MO.4	Organic Waste Management Plan	1.52%	
MO.5	Solid Waste Management Plan	1.90%	

Table 2   Districts Rating System Scope, Part II

DESIGN VERIFICATION

The goal of GSAS Districts Design Verification is to ensure that initial criteria submittals are consistent with the built environment after the construction process is complete. Thus verification will take place in two phases – Initial Review and Final Review.

During Initial Review, projects assessed under GSAS Districts that achieve at least a 1 Star rating will be issued a provisional certificate. The provisional certificate enables the project to apply for a permit to begin the construction process. The provisional certificate does not designate the project as a GSAS certified district.

After construction is complete, projects must resubmit certain requirements for Final Review. For criteria where simulations are required during Initial Review, actual measurements of the existing built environment must be provided for Final Review. The project must resubmit any criteria that may have changed after Preliminary Certification. Additionally, resubmit the following:

- All criteria within the Materials category
- [S.1] Land Preservation
- [S.2] Water Body Preservation
- [S.3] Habitat Preservation
- [S.4] Landscape Amenities
- [S.5] Walkability
- [S.11] Public Space
- [OE.1] Heat Island Effect
- [OE.4] Acoustic Quality
- [CE.2] Support of National Economy
- [MO.2] Management Plan

For any requirements not adequately met during Final Review, the credit for that criterion will be revoked and the score will be recalculated. Based on results of the Final Review, the project will be given a final score and issued a GSAS certification if the constructed project meets or exceeds minimum requirements.

INSTRUCTIONS

GSAS Districts is a rating system for assessing the ecological impacts of district and urban development projects. The rating system is comprised of this manual and an accompanying GSAS Districts Guidelines to describe best practices associated with completing the assessment system. GSAS Districts is divided into eight categories that define the ways a project can impact the environment. The categories are Urban Connectivity [UC], Site [S], Energy [E], Water [W], Materials [M], Outdoor Environment [OE], Cultural & Economic Value [CE] and Management & Operations [MO]. Each category measures a different aspect of the project's environmental impact and addresses ways in which a project can mitigate the negative environmental effects. The categories are then broken down into specific criteria that measure and define individual issues. The issues range from a thorough review of water consumption to an assessment of walkability. Each criterion specifies a process for measuring individual aspects of environmental impact and for documenting the degree to which the requirements have been met. A score is then awarded to each criterion based on the degree of compliance.

Each of the criteria in the assessment system contains the following elements:

DESCRIPTION Outlines the intent of the criterion.

MEASUREMENT PRINCIPLE Summarizes the overall principle of how the criterion will be measured.

MEASUREMENT Describes, in detail, the steps and requirements the project must take in order to demonstrate criterion compliance. Additionally, for certain criteria that require complex computation, tools are provided to facilitate calculations.

SUBMITTAL Provides information on computation or documentation requirements that the project needs to submit in order to demonstrate compliance. These include plans, drawings, simulations, specifications, reports or calculations.

SCORE Lists the range of possible compliance levels and the score associated with each level. Calculated values should be rounded to the nearest value presented within the ranges. For a criterion where the final score is based on the average of two or more factors, if the average is not a whole number, it should be rounded to the nearest whole number. For example: Score a = 1; score b = 2; average = (1 + 2) / 2 = 1.5 which should be rounded to 2.

URBAN CONNECTIVITY [UC]   The Urban Connectivity category consists of factors associated with the urban environment such as zoning, transportation networks and loadings. Loadings on the urban environment include traffic congestion, pollution, external noise, and waste/sewage infrastructure.

IMPACTS   Environmental impacts resulting from unsustainable urban practices include:

- Climate Change
- Fossil Fuel Depletion
- Water Depletion
- Materials Depletion
- Land Use & Contamination
- Water Pollution
- Air Pollution
- Human Comfort & Health

MITIGATE IMPACT   Factors that could mitigate environmental impact include:

- Minimizing the load on the traffic/transportation infrastructure
- Encouraging site selection near existing urban areas to ensure proximity to infrastructure
- Encouraging site selection away from noise producing elements
- Reducing the amount of waste leaving the site

CATEGORY WEIGHT   10%

CRITERIA INCLUDED

No	Criteria	Min Score	Max Score
UC.1	Transportation Load	0	3
UC.2	Proximity to Existing Districts	0	3
UC.3	Acoustic Conditions	0	3
UC.4	Solid Waste Load	0	3
Total Possible		**0**	**12**

URBAN CONNECTIVITY [UC.1] Transportation Load

DESCRIPTION Minimize the transportation load generated by the development in order to reduce vehicle emissions per capita.

MEASUREMENT PRINCIPLE All projects will evaluate the total emission load generated by the district from vehicular transportation.

MEASUREMENT All projects will identify a Transportation Planner or Engineer and will provide a Transportation Demand Model (TDM) to determine the percent reduction in vehicle emissions from unmitigated conditions to mitigated conditions. Mitigated conditions are the factors used to lower vehicle miles traveled within the new development. The TDM can be completed using either a software program or spreadsheet. The TDM tool should include a list of comprehensive inputs and be able to calculate the percent reduction of Vehicle Miles Traveled (VMT).

Minimum inputs for a comprehensive TDM include, but are not limited to:

- Population Variables
- Trip Generation Variables
- Trip Productions
- Trip Attractions
- Total Trip Generations
- Land Use Density and Mix
- Mode Split
- Transit Service Quality
- Freight and Commercial Transport Efficiency
- Multi-Modal Level-of-Service Indicators (Walking, Cycling, Automobile, Public Transit, Taxi, Aviation, Rail)

SUBMITTAL Submit the Transportation Demand Model and the following supporting documents:

- Model input and reference data
- Model output/results

DISTRICTS ASSESSMENT

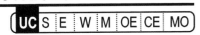

SCORE

Score	% Reduction in Vehicle Emissions (X)
0	$X < 30\%$
1	$30\% \leq X < 40\%$
2	$40\% \leq X < 50\%$
3	$X \geq 50\%$

DISTRICTS ASSESSMENT

URBAN CONNECTIVITY [UC.2] Proximity to Existing Districts

DESCRIPTION Encourage development near existing urban areas to maximize shared use of infrastructure.

MEASUREMENT PRINCIPLE All projects will be located in proximity to developed areas.

MEASUREMENT All projects will complete the Proximity to Existing Districts Calculator and identify on a site map all of the developed, undeveloped, and non-developable land parcels that are located within 1 km of the site boundary. The following definitions apply to the various land parcels:

- Developed land is considered as any parcel of land that has made commercial, industrial, institutional, residential, or landscape improvements to more than 75% of the site. These improvements include but are not limited to apartment buildings, office parks, schools, structured parking, and city parks.
- Undeveloped land is considered as any parcel of land that has improvements on less than 75% of the site. These parcels include, but are not limited to, vacant industrial lots, speculative development, surface parking, and barren land.
- Non-developable land is considered as any parcel of land that cannot be developed or is protected from development. These parcels include, but are not limited to, water bodies, streets, or nature preserves.

The Proximity to Existing Districts Calculator computes the criterion score based on the percent of developed land out of the total developable land within 1 km of the site boundary. The total developable land includes all land not considered as non-developable, and is determined as follows:

Total Developable Land = Developed Land + Undeveloped Land - Non Developable Land

DISTRICTS ASSESSMENT

SUBMITTAL Submit the Proximity to Existing Districts Calculator and a site plan clearly identifying the project boundary and all developed, undeveloped, and non-developable parcels of land within 1 km of the site boundary.

SCORE

Score	% of Developed Area (X)
0	X < 45%
1	45% ≤ X < 60%
2	60% ≤ X < 75%
3	X ≥ 75%

URBAN CONNECTIVITY [UC.3] Acoustic Conditions

DESCRIPTION Encourage selection and design of the site with the least amount of noise exposure.

MEASUREMENT PRINCIPLE All projects will measure the day and night noise level exposure of the district's zones to nearby major noise producing elements.

MEASUREMENT All projects will complete the Acoustic Conditions Calculator to determine the noise level exposures, in decibels (dBA), of the district's zones to nearby noise producing elements. Noise producing elements include, but are not limited to the following:

- Airports
- Highways
- Train Stations
- Industrial Plants

Identify on the district zoning map the location of each of the four zone types: Industrial, Commercial, Residential, and Quiet. Quiet zones are defined as hospital facilities.

All projects will identify all noise producing elements within a reasonable distance from the site boundary, which may pose a negative impact on the site, and all designated zones within the development closest to each source. Take day (peak time) and night (22:00 - 7:00) noise level measurements at the point of the designated zone that is the shortest distance from the edge of the noise source. These noise levels should not exceed the following levels:

Zone Types	L_{day}	L_{night}
Industrial	70 dBA	65 dBA
Commercial	65 dBA	55 dBA
Residential	55 dBA	45 dBA
Quiet	45 dBA	35 dBA

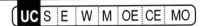
SUBMITTAL Submit the Acoustic Conditions Calculator and the following supporting documents:

- Plan identifying all noise producing elements, the affected zones, and the measurement points for each zone
- List of each measurement point and the time that each day and night measurement was taken

SCORE

Score	Requirement
0	L_{day} and L_{night} exceed the allowed maximum noise levels in one or more zones.
3	L_{day} and L_{night} are within the allowed maximum noise levels for all zones.

URBAN CONNECTIVITY [UC.4] Solid Waste Load

DESCRIPTION Minimize solid waste load by recycling and composting to reduce the burden on public facilities.

MEASUREMENT PRINCIPLE All projects will minimize the amount of solid waste going into municipal infrastructure, such as landfills and incineration facilities.

MEASUREMENT All projects will complete the Solid Waste Load Calculator to compute the percent of solid waste not being recycled or composted, using the following variables:

- Total Composting Capacity (tonnes/year)
- Total Recycling Capacity, if applicable (tonnes/year)
- Total Solid Waste Generated (tonnes/year)

It is recommended that recycling and composting facilities be built on-site to handle the new district solid waste load. For smaller developments that are part of an existing district framework, shared waste infrastructure is acceptable with proof of the facility's capacity and district's collection system.

SUBMITTAL Submit the Solid Waste Load Calculator and any supporting documents that demonstrate solid waste production and the capacity to compost or recycle on-site. Supporting documents may include:

- Location of composting or recycling facilities and specifications
- Drawings indicating a collection system

SCORE

Score	% of Solid Waste Not Recycled(X)
0	X > 50%
1	40% < X ≤ 50%
2	30% < X ≤ 40%
3	X ≤ 30%

SITE [S] The Site category consists of factors associated with land use such as land conservation or remediation, site selection, planning and development.

IMPACTS Environmental impacts resulting from improper land use and unsustainable practices include:

- Climate Change
- Fossil Fuel Depletion
- Water Depletion
- Materials Depletion
- Land Use & Contamination
- Water Pollution
- Air Pollution
- Human Comfort & Health

MITIGATE IMPACT Factors that could mitigate environmental impact due to land use include:

- Selecting a site that has minimal ecological value or is contaminated
- Preserving natural water bodies on or nearby the site
- Preserving habitats that exist on the site
- Defining a landscaping plan that encourages the use of native vegetation
- Creating pedestrian and bicycle pathways to reduce travel and enhance mobility
- Preventing and reversing desertification
- Reducing the parking footprint to minimize parking surfaces
- Designing for mixed uses to reduce travel
- Developing strategies to prevent crime on-site
- Creating public spaces to encourage social interaction and promote physical and mental well-being
- Constructing GSAS rated buildings to ensure sustainable practices at all scales of the district

CATEGORY WEIGHT 20%

DISTRICTS ASSESSMENT

UC **S** E W M OE CE MO

CRITERIA INCLUDED

No	Criteria	Min Score	Max Score
S.1	Land Preservation	-1	3
S.2	Water Body Preservation	-1	3
S.3	Habitat Preservation	-1	3
S.4	Vegetation	-1	3
S.5	Walkability	-1	3
S.6	Bikeability	-1	3
S.7	Desertification	-1	3
S.8	Parking Footprint	-1	3
S.9	Mixed Use	-1	3
S.10	Crime Prevention	-1	3
S.11	Public Space	-1	3
S.12	GSAS Rated Typologies	-1	3
Total Possible		**-12**	**36**

SITE [S.1] Land Preservation

DESCRIPTION Encourage development on land that is contaminated, previously developed or has low ecological value. In addition, preserve or enhance the site through remediation, conservation, and/or restoration.

MEASUREMENT PRINCIPLE All projects will assess the site for its soil quality and contamination level and determine strategies to conserve, restore, or enhance the site.

MEASUREMENT An Ecologist or Land Specialist will assess the site for all projects in order to complete the Land Preservation Calculator and to create a Site Assessment Report that identifies the following:

- Soil Quality (low, moderate, high)
- Areas of the site that are contaminated or have been previously developed
- Options for conserving the site in its natural state
- Recommendations for remediating contaminated land or restoring/enhancing barren areas of the site
- Additional requirements for particularly sensitive sites

Additionally, if the site is contaminated, the ecologist will identify the following:

- Contaminant sources/types
- Degree of contamination
- Strategies for remediating the sources of contamination

All projects will also develop a Soil Erosion Plan to identify the potential for soil erosion as well as to provide recommendations for acceptable ways to limit soil erosion during and after construction. The Soil Erosion Plan should follow the standards of the ASTM Subcommittee D18.25 on Erosion and Sediment Control Technology and consist of a written report with all associated addenda (maps, figures, references, and backup calculations), specifically to include the following:

MEASUREMENT

- Site map and written description identifying soil texture and topography. Outline any areas that are within risk zones for flooding, runoff, or any other type of water erosion, if applicable.
- Hydraulic and hydrologic data for the site.
- Description of site drainage features, drainage characteristics, and drainage calculations for the site after completion of the project. This should include the influence and location of site groundwater levels, if applicable.
- Erosion control calculations for both short-term (during construction) and long-term (after project completion) surface conditions.
- Stability of earthen embankments, if necessary, including backup calculations. This should also include recommendations for maintaining excavation stability (sloping, benching, or shoring requirements) in light of any potential water erosion hazards identified at the site.
- Descriptions and locations of all existing and proposed on-site drainage facilities.
- Timeline sequence of each proposed earth change (grading and earthwork) and time exposure of each area prior to the completion of effective soil erosion and sediment control measures.
- Descriptions and locations of all recommended proposed temporary and permanent soil erosion and sedimentation control measures. Alternative permanent control measures should be presented along with advantages and disadvantages of each measure for review.
- Recommendations for gradation requirements of soils to be used, including recommendations of types of compacted fills and surficial soil cover at the site with consideration for potential water erosion.
- Provisions for sediment control during construction, including tracking of sediment off-site by construction vehicles.
- Program outlining the continued maintenance of all permanent soil erosion measures and sedimentation control facilities.

SUBMITTAL Submit the Land Preservation Calculator and the following supporting documents:

- Site Assessment Report
- Soil Erosion Plan
- Site plan identifying areas of varying degrees of soil quality, soil disturbance, and contaminated land for pre- and post-development
- Site plan identifying areas to be conserved, restored, or enhanced
- Specifications that illustrate how site remediation, conservation, and restoration outlined in the Site Assessment Report will be implemented
- Any other drawings that meet the specific requirements of the Site Assessment Report

SCORE If either factor (a) or (b) receives a -1, the final score for the project is a -1. Otherwise, the final score is the average of (a) and (b).

Score	Performance Indicator (a)
-1	$a < 0.00$
0	$0.00 \leq a < 0.25$
1	$0.25 \leq a < 0.50$
2	$0.50 \leq a < 0.75$
3	$a \geq 0.75$

Score	Soil Erosion Plan (b)
-1	Soil Erosion Plan does not demonstrate compliance.
3	Soil Erosion Plan demonstrates compliance.

SITE [S.2] Water Body Preservation

DESCRIPTION Encourage development that prevents or minimizes ecological degradation to water bodies in order to preserve the natural resources of the region.

MEASUREMENT PRINCIPLE All projects will develop strategies to conserve, restore, or enhance the water bodies on or nearby the site.

MEASUREMENT An Ecologist or Land Specialist will evaluate the site for all projects in order to complete a Water Body Preservation Plan which will contain strategies and guidelines for the conservation, restoration, and/or enhancement of natural water bodies on or nearby the project site. Water bodies are defined as areas that hold surface or ground water, including, but not limited to, streams, rivers, lakes, estuaries, bays, gulfs, and aquifers. The Water Body Preservation Plan must address the following requirements:

- Water Body Conservation: Demonstrate that the project will preserve all existing water bodies. There is no acceptable limit for infilling or polluting of existing water bodies.
- Coastal Protection: Show how the project will protect ALL nearby coastlines from damage or pollution as a result of the development.
- Groundwater Protection: Demonstrate that the project will prevent damage or contamination of groundwater.

SUBMITTAL Submit the Water Body Preservation Plan and the following supporting documents:

- Diagrams, drawings, or plans identifying the location of natural water bodies pre- and post-development
- Diagrams or drawings indicating at least a 200m buffer around coastal area
- Documents outlining the steps necessary for preventing groundwater contamination during the operations phase of the development

SCORE

Score	Requirement
-1	Water Body Preservation Plan does not demonstrate compliance.
3	Water Body Preservation Plan demonstrates compliance.

SITE [S.3] Habitat Preservation

DESCRIPTION Encourage development that preserves and/or enhances the biodiversity of the site in order to protect the natural ecosystems of the region.

MEASUREMENT PRINCIPLE All projects will develop strategies to conserve, restore, or enhance habitats and the biodiversity of the site.

MEASUREMENT An Ecologist or Land Specialist will evaluate the site for all projects and create a Habitat Preservation Plan that identifies the following:

- All habitats within the site and on adjacent areas. Examples of habitats include mangroves, wadis, and deserts. Habitats can also be defined by species of animals and vegetation.
- Ecologically sensitive habitats which house endangered species of plants or animals.
- Strategies for protecting all endangered species and their habitats.

All projects will identify and protect all habitats that are specified as endangered in the Habitat Preservation Plan. The Habitat Preservation Plan will be scored based on the quality of the habitats maintained on the site.

SUBMITTAL Submit the Habitat Preservation Plan and the following supporting documents:

- Diagrams or drawings that identify all habitats, pre- and post-development
- List of endangered plant and animal species
- Diagrams, drawings, or plans that illustrate strategies for preserving ecosystem interaction within the site and adjacent areas

UC **S** E W M OE CE MO

SCORE

Score	Requirement
-1	Habitat Preservation Plan does not demonstrate compliance.
3	Habitat Preservation Plan demonstrates compliance.

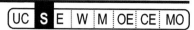

DISTRICTS ASSESSMENT

UC | S | E | W | M | OE | CE | MO

SITE [S.4] Vegetation

DESCRIPTION Minimize lawn and encourage native or low-impact vegetation for the site in order to reduce irrigation demand.

MEASUREMENT PRINCIPLE All projects will develop a landscaping plan to minimize the amount of lawn and increase the amount of native vegetation or low-impact vegetation.

MEASUREMENT All projects will complete the Vegetation Calculator to assess the landscaping plan. The project will meet the following requirements:

- The lawn must not exceed 50% of the total landscaped area, excluding roadways and hardscaped areas. Where appropriate, use vegetation that minimizes water usage.
- At least 30% of all vegetation planted within the site should be native to the climate of the region. Refer to the list of Recommended Plant Species in the *GSAS Design Guidelines.*

SUBMITTAL Submit the Vegetation Calculator and the following supporting documents:

- Landscape plan and planting schedule identifying the type and location of plantings
- Landscape material data sheets

Districts: GSAS Assessment v2.0 - 2013

PAGE 36

DISTRICTS ASSESSMENT

SCORE If either factor (a) or (b) receives a -1, the final score for the project is a -1. Otherwise, the final score is the average of (a) and (b).

Score	% of Lawn (a)
-1	a > 50%
0	30% < a ≤ 50%
1	15% < a ≤ 30%
2	0% < a ≤ 15%
3	a = 0%

Score	% of Native Vegetation (b)
-1	b < 30%
0	30% ≤ b < 45%
1	45% ≤ b < 60%
2	60% ≤ b < 75%
3	b ≥ 75%

SITE [S.5] Walkability

DESCRIPTION Encourage sustainable infrastructure through development of efficient, user-friendly pedestrian pathways.

MEASUREMENT PRINCIPLE All projects will determine the extent and usability of pedestrian pathways.

MEASUREMENT All projects will complete the Walkability Calculator to determine the extent and usability of the projects pedestrian pathways. The project will determine the ratio of pedestrian pathway length (meters) to vehicular roadway length (meters) to evaluate the extent of pathways within the development. Additionally, the project will perform a shading simulation to compute the percent of applicable pedestrian pathways shaded by both landscape features and buildings in order to determine the usability of pathways within the development.

Pedestrian pathways may occur along roadways, within public spaces, or between buildings. For the purposes of this criterion, projects will only measure hardscaped pedestrian pathways within the development. The following road types are defined by AASHTO (American Association of State Highway and Transportation Officials):

- Arterial: Provides the highest level of service at the greatest speed for the longest uninterrupted distance, with some degree of access control.
- Collector: Provides a less highly developed level of service at a lower speed for shorter distances by collecting traffic from local roads and connecting them with arterials
- Local: Consists of all roads not defined as arterials or collectors; primarily provides access to land with little or no through movement.

For this criterion, only collector and local roads should be considered when determining the total length of roadways within the development.

SUBMITTAL Submit the Walkability Calculator and the following supporting documents:

- Site plan demonstrating the length of pedestrian pathways and roadways identified as arterial, collector, or local roads
- Landscape plan and planting schedule identifying the type and location of plantings and the type and location of architectural shading features
- Landscape material data sheets
- Results of the shading simulation performed on June 21st at 15:00 that demonstrate the areas of the pedestrian pathways that are shaded

SCORE If either factor (a) or (b) receives a -1, the final score for the project is a -1. Otherwise, the final score is the average of (a) and (b).

Score	Ratio of Pedestrian Pathway Length to Vehicular Roadway Length (a)
-1	a < 1.25
0	1.25 ≤ a < 1.50
1	1.50 ≤ a < 1.75
2	1.75 ≤ a < 2.00
3	a ≥ 2.00

Score	% of Pedestrian Pathways Shaded (b)
-1	b < 60%
0	60% ≤ b < 70%
1	70% ≤ b < 80%
2	80% ≤ b < 90%
3	b ≥ 90%

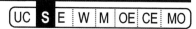
SITE [S.6] Bikeability

DESCRIPTION Encourage sustainable infrastructure through the development of efficient, user-friendly bicycle pathways.

MEASUREMENT PRINCIPLE All projects will determine the extent and usability of bicycle pathways.

MEASUREMENT All projects will complete the Bikeability Calculator to determine the extent of bicycle pathways. Bicycle pathways may occur along roadways as bike lanes or as separated bike paths through public spaces. Pedestrian pathways may not be counted as bicycle pathways.

The Bikeability Calculator computes the criterion score based on the ratio of bicycle pathway length (meters) to total vehicular roadway length (meters). Bicycle pathways are separate from pedestrian pathways and may occur along roadways, within public spaces, or between buildings. The following road types are defined by AASHTO (American Association of State Highway and Transportation Officials):

- Arterial: Provides the highest level of service at the greatest speed for the longest uninterrupted distance, with some degree of access control.
- Collector: Provides a less highly developed level of service at a lower speed for shorter distances by collecting traffic from local roads and connecting them with arterials.
- Local: Consists of all roads not defined as arterials or collectors; primarily provides access to land with little or no through movement.

SUBMITTAL Submit the Bikeability Calculator and a site plan demonstrating the length of bicycle pathways and roadways identified as arterial, collector, or local roads.

DISTRICTS ASSESSMENT

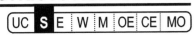

SCORE

Score	Ratio of Bicycle Pathway Length to Vehicular Roadway Length (X)
-1	$X < 0.075$
0	$0.075 \leq X < 0.100$
1	$0.100 \leq X < 0.125$
2	$0.125 \leq X < 0.150$
3	$X \geq 0.150$

SITE [S.7] Desertification

DESCRIPTION Reverse, prevent, or minimize desertification and protect the development site from sandstorms.

MEASUREMENT PRINCIPLE All projects will prevent desertification by protecting the development's landscape from sandstorms and other adverse wind conditions.

MEASUREMENT All projects will provide a Desertification Plan, reviewed by an Ecologist, that outlines physical strategies for reversing, preventing, or minimizing desertification at the district scale.

SUBMITTAL Submit a Desertification Plan and the following supporting documents:

- Site plan indicating the location of any physical desertification strategies
- Diagrams demonstrating the scope and specification of desertification strategies

SCORE

Score	Requirement
-1	Desertification Plan does not demonstrate compliance.
3	Desertification Plan demonstrates compliance.

SITE [S.8] Parking Footprint

DESCRIPTION Minimize the parking footprint within the development to reduce the amount of impervious surface parking on the site.

MEASUREMENT PRINCIPLE Project should minimize the total footprint dedicated to parking as a percentage of the total development area.

MEASUREMENT All projects will complete the Parking Footprint Calculator to determine the parking footprint compared to the total footprint of the development. The following parking types should be included in the calculator:

- Surface parking lots
- Parking structure footprint, where the entirety of the structure is intended for parking
- Alternative paving solutions for surface lot parking, including porous concrete, porous asphalt, or permeable pavers

The following parking types are excluded from the calculation:

- On-street, parallel parking
- Underground parking
- Parking structure footprint, where there are multiple uses included in the structure (i.e. residential, retail, office)

SUBMITTAL Submit the Parking Footprint Calculator and the following supporting documents:

- Site plan identifying all surface parking lots and parking structures
- Material data sheets for all pervious or alternative paving systems used

SCORE

Score	% of Development Area Dedicated to Parking (X)
-1	X > 20%
0	15% < X ≤ 20%
1	10% < X ≤ 15%
2	5% < X ≤ 10%
3	X ≤ 5%

SITE [S.9] Mixed Use

DESCRIPTION Maximize the number of major uses within the development in order to reduce the need for transport.

MEASUREMENT PRINCIPLE All projects will maximize the number of major uses within the development.

MEASUREMENT All projects will complete the Mixed Use Calculator to determine the number of diverse uses within walking distance to the proposed population.

Uses are defined by the following categories:

Population

- Residential
- Office Space

Use Categories

- Public Services
- Places of Worship
- Retail - Services
- Retail - Goods
- Retail - Food

The Mixed Use Calculator computes the criterion score based on the population and number of uses within each 480 meter grid block, delineated by the project on a site plan. The number of uses is weighted by the population in order to calculate a Performance Indicator.

SUBMITTAL Submit the Mixed Use Calculator and a site plan identifying the location of planned amenities and the population within each of the 480 meter grid blocks.

SCORE

Score	Performance Indicator (X)
-1	$X < 1.00$
0	$1.00 \leq X < 2.00$
1	$2.00 \leq X < 3.00$
2	$3.00 \leq X < 4.00$
3	$X \geq 4.00$

SITE [S.10] Crime Prevention

DESCRIPTION Encourage safety measures in the design of the built environment to deter crime.

MEASUREMENT PRINCIPLE All projects will develop a built environment that is designed to deter crime.

MEASUREMENT All projects will create a Crime Prevention Plan showing the intent of safe lighting levels in all areas of the district. The plan should meet, but not significantly exceed, the required IESNA light levels for the following categories:

- Security Lighting
- Roadway/Street Lighting
- Walkway/Bikeway Lighting
- Pedestrian Mall, Plaza, and Park Lighting
- Parking Lot Lighting

Reference IESNA RP-33-99, RP-8, and RP-20 for recommended light levels.

SUBMITTAL Submit the Crime Prevention Plan and the intended light levels in all shared public spaces within the district.

SCORE

Score	Requirement
-1	Crime Prevention Plan does not demonstrate compliance.
3	Crime Prevention Plan demonstrates compliance.

SITE [S.11] Public Space

DESCRIPTION Encourage social interaction and promote the physical and mental well-being of the community by providing accessible and usable outdoor public space.

MEASUREMENT PRINCIPLE All projects will provide an adequate amount of public space for the district users, as well as ensure that the public spaces are easily accessible and include appropriate levels of shading.

MEASUREMENT All projects will complete the Public Space Calculator to determine the accessibility, area, and usability of public space. Public spaces can include parks, plazas, recreational facilities, sports fields, community facilities, and other spaces that are open and accessible to the general public.

The calculator computes a Public Space Performance Indicator based on the public space per capita and the accessibility of public spaces to the development users. Additionally, the project will perform a shading simulation to determine the percent of public spaces shaded.

SUBMITTAL Submit the Public Space Calculator and the following supporting documents:

- Plan showing the location and size of all public spaces. Identify a 480 meter boundary surrounding each public space and the buildings within that boundary
- Report and/or drawing documenting the total number of users in the development and the number of users that fall within the 480 meter boundary of each public space
- Landscape plan and planting schedule identifying the type and location of plantings and the type and location of architectural shading features
- Landscape material data sheets
- Results of the shading simulation performed on June 21st at 15:00 that demonstrate the areas of public spaces that are shaded

SCORE If either factor (a) or (b) receives a -1, the final score for the project is a -1. Otherwise, the final score is the average of (a) and (b).

Score	Public Space Performance Indicator (a)
-1	a < 15
0	15 ≤ a < 20
1	20 ≤ a < 25
2	25 ≤ a < 30
3	a ≥ 30

Score	% of Public Spaces Shaded (b)
-1	b < 25%
0	25% ≤ b < 30%
1	30% ≤ b < 35%
2	35% ≤ b < 40%
3	b ≥ 40%

DISTRICTS ASSESSMENT

UC S E W M OE CE MO

SITE   [S.12] GSAS Rated Typologies

DESCRIPTION   Encourage sustainability through the rating of individual buildings.

MEASUREMENT PRINCIPLE   All projects will assess whether GSAS is required at the building level and the average score for the district.

MEASUREMENT   All projects will submit the GSAS Rated Typologies Calculator to determine the average GSAS Design score for the district. The score for each GSAS Design rated building is weighted by its gross floor area (m²)

SUBMITTAL   Submit the GSAS Rated Typologies Calculator and a site plan identifying all GSAS rated buildings with individual scores and corresponding gross floor areas.

SCORE

Score	Weighted Average GSAS Typology Rating (X)
-1	X < -0.3 OR less than 70% of District requires GSAS Design
0	$-0.3 \leq X < 0.5$
1	$0.5 \leq X < 1.5$
2	$1.5 \leq X < 2.5$
3	$X \geq 2.5$

DISTRICTS ASSESSMENT

ENERGY [E]   The Energy category consists of factors associated with the efficiency of energy delivery and the use of fossil energy sources that result in harmful emissions and pollution.

IMPACTS   Negative impacts resulting from energy use and unsustainable practices include:

- Climate Change
- Fossil Fuel Depletion
- Air Pollution
- Human Comfort & Health

MITIGATE IMPACT   Factors that could mitigate environmental impacts due to energy use include:

- Selecting efficient building systems
- Lowering the demand on nonrenewable sources of energy, thereby reducing the depletion of fossil fuels
- Reducing harmful emissions
- Minimizing the amount of harmful substances produced by the energy delivery systems and the energy supply network

CATEGORY WEIGHT   18%

CRITERIA INCLUDED

No	Criteria	Min Score	Max Score
E.2	Energy Delivery Performance	-1	3
E.3	Fossil Fuel Depletion	-1	3
E.4	CO_2 Emissions	-1	3
E.5	NO_x, SO_x, & Particulate Matter	-1	3
Total Possible		**-4**	**12**

DISTRICTS ASSESSMENT

UC | S | **E** | W | M | OE | CE | MO

ENERGY [E.2] Energy Delivery Performance

DESCRIPTION Establish energy delivery performance of all systems that serve the district.

MEASUREMENT PRINCIPLE All projects will conduct assessment of integrated district energy performance in relation to the baseline and targets outlined in the District Energy Performance Calculator.

MEASUREMENT All projects will complete the District Energy Performance Calculator to determine the District's Delivered Energy Performance Coefficient (EPC_{del}). The district's energy delivery performance (E_{del}) is based on several input parameters including:

1. Built objects or planned parks that are or will be GSAS rated:

- Total floor/site area (m^2) per typology
- GSAS target score (0,1,2,3)* for [E.1] Energy Demand Performance
- GSAS target score (0,1,2,3)* for [E.2] Energy Delivery Performance

*Note: There is a predefined range of Demand Energy Performance Coefficients (EPC_{nd}) and Delivered Energy Performance Coefficients (EPC_{del}) that corresponds to any given target score for a building in the GSAS rating system. The middle point of the specified range will be assigned for the EPC_{nd} and EPC_{del} values to compute the energy intensity based on the user input target score.

2. Built objects that currently cannot be GSAS rated:

- Total floor/site area (m^2) per typology
- Target energy improvement percentage that indicates performance relative to the reference values $Q_{nd,ref}$ and E_{ref_del}**

**Note: Reference values for GSAS non-rated building typologies will be further developed and provided by GSAS. Any building typologies for which GSAS does not provide reference values are omitted from district energy assessment.

MEASUREMENT   3. Infrastructure Delivered Energy

Option 1

- Water supply energy performance
- Wastewater treatment energy performance
- District cooling plant pump energy performance
- Irrigation energy performance
- Park lighting energy performance
- Traffic lighting energy performance
- Street lighting energy performance
- Other energy consuming components

Option 2

- Overall infrastructure energy intensity for non-built areas

4. District Energy System

- Percentage provided by district cooling plant to meet the annual cooling demand of the project

5. Renewable Energy Generation

- Percentage of district-scale energy generation from power station using renewable resources such as solar, wind, geothermal or biomass

Calculated $EPC_{del} = E_{del} / E_{ref\_del}$

With $E_{ref\_del} = \dfrac{\sum A_i\, E_{(ref\_del,i)} + A_{non\text{-}built}\, E_{ref\_infra}}{\sum A_i + A_{non\text{-}built}}$

Where:

- E_{del} is calculated according to the District Energy Performance Calculator.
- i represents an individual building type, including GSAS rated as well as non-rated buildings.
- A_i represents building floor area for the building type.
- $A_{non\text{-}built}$ represents all areas outside the building footprint including parks, open spaces (public or private), streets, parking lots, etc.

SUBMITTAL Submit the District Energy Performance Calculator and the following supporting documents:

- District design drawings
- Target scores for all GSAS rated buildings. In the case of already designed buildings with GSAS certification, use actual scores achieved for [E.1] and [E.2]
- Target energy improvements for buildings not rated by GSAS
- Documentation showing buildings and objects that have been excluded from the District Energy Performance Calculator
- Documentation identifying energy performance inputs for infrastructure energy use, district cooling efficiency, and percentage of energy supplied by renewable resources

SCORE

Score	EPC_{del} Value
-1	EPC > 1.0
0	0.8 < EPC ≤ 1.0
1	0.7 < EPC ≤ 0.8
2	0.6 < EPC ≤ 0.7
3	EPC ≤ 0.6

ENERGY [E.3] Fossil Fuel Conservation

DESCRIPTION Establish fossil energy conservation performance of the delivery systems and energy supply network that serve the district.

MEASUREMENT PRINCIPLE All projects will conduct assessment of district fossil fuel performance in relation to the baseline and targets outlined in the District Energy Performance Calculator.

MEASUREMENT All projects will complete the District Energy Performance Calculator to determine the Primary Energy Performance Coefficient (EPC_p). The district's primary energy performance (E_p) value is determined using the District Energy Performance Calculator, from the calculated energy delivery performance and selected energy carrier percentage to meet the total energy consumption from power plant.

Calculated $EPC_p = E_p / E_{ref\_p}$

Where:

- E_p is calculated according to the District Energy Performance Calculator.
- E_{ref_p} is determined by the District Energy Performance Calculator on the basis from the calculated E_{ref_del} and primary energy factor.

SUBMITTAL Submit the District Energy Performance Calculator and the following supporting documents:

- District power distribution plan and drawings
- Energy supply network plan from power plant to district for each energy carrier used

SCORE

Score	EPC_p Value
-1	EPC > 1.0
0	0.8 < EPC ≤ 1.0
1	0.7 < EPC ≤ 0.8
2	0.6 < EPC ≤ 0.7
3	EPC ≤ 0.6

ENERGY [E.4] CO$_2$ Emissions

DESCRIPTION Establish CO$_2$ emission reduction performance of the delivery systems and energy supply network that serve the district.

MEASUREMENT PRINCIPLE All projects will conduct assessment of the district's CO$_2$ emission performance in relation to the baseline and targets outlined in the District Energy Performance Calculator.

MEASUREMENT All projects will complete the District Energy Performance Calculator to determine the District's CO$_2$ Emissions Performance Coefficient (EPC$_{CO2}$). District CO$_2$ emission performance (CO$_2$) is determined using the District Energy Performance Calculator from the calculated energy delivery performance and selected energy carrier percentage to meet the total energy consumption from power plant.

Calculated $EPC_{CO2} = CO_2 / CO_{2,ref}$

Where:

- CO$_2$ emissions are calculated according to the District Energy Performance Calculator.
- CO$_{2,ref}$ is determined by the District Energy Performance Calculator on the basis of the calculated E_{ref_del} and CO$_2$ emission coefficient.

SUBMITTAL Submit the District Energy Performance Calculator and the following supporting documents:

- District power distribution plan and drawings
- Energy supply network plan from power plant to district for each energy carrier used

SCORE

Score	EPC_{CO2} Value
-1	EPC > 1.0
0	0.8 < EPC ≤ 1.0
1	0.7 < EPC ≤ 0.8
2	0.6 < EPC ≤ 0.7
3	EPC ≤ 0.6

ENERGY [E.5] NO$_x$, SO$_x$, & Particulate Matter

DESCRIPTION Establish NO$_x$, SO$_x$, and dust emission reduction performance of the district with its delivery systems and energy supply network.

MEASUREMENT PRINCIPLE All projects will conduct assessment of district NO$_x$, SO$_x$, and dust emission performance in relation to the baseline and targets outlined in the District Energy Performance Calculator.

MEASUREMENT All projects will complete the District Energy Performance Calculator to determine the District's NO$_x$, SO$_x$, and Dust Emission Performance Coefficient (EPC$_{NOx-SOx}$). District NO$_x$, SO$_x$, and dust emission performance (NO$_x$ and SO$_x$) values are determined using the District Energy Performance Calculator from the calculated energy delivery performance and selected energy carrier percentage to meet the total energy consumption from power plant.

Calculated EPC$_{NOx-SOx}$ = {(NO$_x$ / NO$_{x.ref}$) + (SO$_x$/ SO$_{x.ref}$)} / 2

Where:

- NO$_x$ and SO$_x$ emissions are calculated according to the District Energy Performance Calculator.
- NO$_{x.ref}$ and SO$_{x.ref}$ are determined by the District Energy Performance Calculator on the basis of the calculated E$_{ref\_del}$ and NO$_x$ / SO$_x$ emission coefficients.

SUBMITTAL Submit the District Energy Performance Calculator and the following supporting documents:

- District power distribution plan and drawings
- Energy supply network plan from power plant to district for each energy carrier used

SCORE

Score	EPC$_{NOx-SOx}$ Value
-1	EPC > 1.0
0	0.8 < EPC ≤ 1.0
1	0.7 < EPC ≤ 0.8
2	0.6 < EPC ≤ 0.7
3	EPC ≤ 0.6

WATER [W]   The Water category consists of factors associated with water consumption and its associated burden on municipal supply and treatment systems.

IMPACTS   Environmental impacts resulting from water consumption and unsustainable practices include:

- Water Depletion
- Human Comfort & Health

MITIGATE IMPACT   Factors that could mitigate environmental impact and lower demand on water include:

- Designing a landscaping plan that minimizes the need for irrigation
- Creating a system for the collection and storage of rainwater
- On-site treatment of water for later reuse

CATEGORY WEIGHT   16%

CRITERIA INCLUDED

No	Criteria	Min Score	Max Score
W.1	Water Consumption	-1	3
Total Possible		**-1**	**3**

WATER   [W.1] Water Consumption

DESCRIPTION Minimize water consumption in order to reduce the burden on municipal supply and treatment systems.

MEASUREMENT PRINCIPLE All projects will conduct an assessment of integrated district water performance in relation to the baseline and targets outlined in the Water Consumption Calculator.

MEASUREMENT All projects will complete the Water Consumption Calculator to determine cumulative water performance for the district. Cumulative water performance is determined by several input parameters including:

- GSAS target scores
- Target water consumption reduction percentage for buildings that cannot be rated with GSAS*
- Project site area
- Landscaping and irrigation plan
- Rainwater and stormwater collection and reuse
- Greywater and blackwater treatment and reuse

*Note: Reference values for GSAS non-rated building typologies will be further developed and provided by GSAS. Any building typologies for which GSAS does not provide reference values are omitted from the district water assessment.

Based on input parameters provided by the project, the calculator computes weighted scores for GSAS rated schemes, non-GSAS rated buildings with reference consumption, and non-GSAS rated sites including landscape areas and water features.

GSAS RATED SCHEMES

The GSAS rated schemes calculated value (Weighted Score$_{GSAS}$) is derived using the input target GSAS design scores for each GSAS rated typology on site.

$$\text{Weighted Score}_{GSAS} = (\text{Target Score}_i \times \text{Site Area}_i) + ... + (\text{Target Score}_n \times \text{Site Area}_n)$$

Where (i-n) represents each project site within the development that has a GSAS rating system score.

NON-GSAS RATED BUILDINGS

The non-GSAS rated buildings calculated value (Weighted Score $_{Non-rated}$) utilizes the target reduction input to calculate a relative target score for each non-rated building typology on-site.

$$\text{Weighted Score}_{Non-rated} = (\text{Target Score}_i \times \text{Site Area}_i) + ... + (\text{Target Score}_n \times \text{Site Area}_n)$$

Where (i-n) represents each non-rated building typology on-site that has a reference consumption value for non-rated buildings.

NON-GSAS RATED SITES

For site areas not related directly to a building or a GSAS rated Park, the non-GSAS rated sites parameters, input by the project, are used to compute the relative efficiency of the landscaping or water feature per square meter of site. The non-rated sites consumption has reference input variables for best and worst case values. A target score is assigned based on the efficiency of the various site areas referencing the best and worst case scenarios. The non-GSAS rated sites value (Weighted Score $_{Non-rated\ sites}$) is then computed using the assigned target scores and total landscaped site area.

$$\text{Weighted Score}_{Non-rated\ sites} = (\text{Target Score} \times \text{Total Landscaped Area})$$

Finally, the Water Consumption Calculator uses the weighted scores for GSAS rated schemes, non-GSAS rated buildings, and non GSAS-rated sites, which are combined and normalized by total site area, to determine the final Weighted Average Score.

$$\text{Weighted Average Score} = \frac{\Sigma \text{Weighted Scores}}{\text{Total Site Area}}$$

SUBMITTAL — Submit the Water Consumption Calculator and the following supporting documents:

- Landscaping and irrigation plan
- Rainwater and stormwater collection and reuse plan
- Greywater and blackwater treatment and reuse plan
- Documentation showing buildings and objects that have been excluded from the Water Consumption Calculator

SCORE

Score	Weighted Average Score (X)
-1	$X < 0$
0	$0 \leq X < 0.5$
1	$0.5 \leq X < 1.5$
2	$1.5 \leq X < 2.5$
3	$X \geq 2.5$

MATERIALS [M] The Materials category consists of factors associated with material extraction, processing, manufacturing, distribution, use/reuse, and disposal.

IMPACTS Environmental impacts resulting from material use and unsustainable practices include:

- Materials Depletion
- Climate Change
- Fossil Fuel Depletion
- Air Pollution
- Human Comfort & Health

MITIGATE IMPACT Factors that could mitigate environmental impact due to material use include:

- Using local materials to reduce transportation needs
- Using responsibly sourced materials
- Using materials with high recycled contents
- Recycling and reusing materials, on- and off-site

CATEGORY WEIGHT 8%

CRITERIA INCLUDED

No	Criteria	Min Score	Max Score
M.1	Regional Materials	-1	3
M.2	Responsible Sourcing of Materials	-1	3
M.3	Recycled Materials	-1	3
M.4	Materials Reuse	-1	3
M.5	Life Cycle Assessment (LCA)	N/A	N/A
Total Possible		**-4**	**12**

MATERIALS [M.1] Regional Materials

DESCRIPTION Encourage the use of regionally manufactured and assembled building elements and materials in order to reduce the carbon footprint of the materials.

MEASUREMENT PRINCIPLE All projects will use materials that are regionally manufactured and assembled.

MEASUREMENT All projects will complete the Regional Materials Calculator to compute a Performance Indicator based on the weight and sourcing distance of all applicable materials. Materials sourcing distance can either be regional (less than or equal to 200 kilometers) or external (over 200 kilometers).

All projects will only consider materials permanently installed in the project. Exclude mechanical, electrical, and plumbing assemblies, as well as specialty items and equipment.

Materials which account for less than 10% by weight or volume of a component may be excluded.

SUBMITTAL For initial review, submit the Regional Materials Calculator and the following supporting documents:

- Intended list of materials
- Documentation outlining the intended manufacturers and sourcing distances

For final review, submit the Regional Materials Calculator and the following supporting documents:

- Bill of Quantities
- List of materials
- Specification listing manufacturers and locations
- Cost estimates
- Report providing distances of manufacturers

SCORE

Score	Performance Indicator (X)
-1	$X \geq 30$
0	$20 \leq X < 30$
1	$10 \leq X < 20$
2	$1 \leq X < 10$
3	$X = 1$

DISTRICTS ASSESSMENT

MATERIALS   [M.2] Responsible Sourcing of Materials

DESCRIPTION   Encourage the use of responsibly sourced materials for primary infrastructure elements in order to minimize the depletion of non-renewable materials.

MEASUREMENT PRINCIPLE   All projects will use materials that are responsibly sourced.

MEASUREMENT   All projects will complete the Responsible Sourcing of Materials Calculator to determine the percent, by cost, of imported materials that can be traced through the supply chain. Applicable materials include all primary building elements that are imported.

All projects will only consider materials permanently installed in the project. Exclude mechanical, electrical, and plumbing assemblies, as well as specialty items and equipment. Materials which account for less than 10% by weight or volume of a component may be excluded.

To ensure the selection of socially and environmentally conscious materials that employ responsible practices throughout the supply chain, all imported materials must comply with the following standards, where applicable, and should be accounted for in the calculator to qualify as responsibly sourced:

Social Responsibility Policies
Materials should follow the standards established by ISO 26000, which defines core issues, principles, and practices relating to integrating, implementing, and promoting socially responsible behavior in organizational practices.

Quality Assurance & Supply Chain Management Systems
Materials should originate from sources with a documented quality management system based on the standards of ISO 9001, and material purchasing should follow the guidelines established by Sub-clause 7.4 of ISO 9001. If accredited systems already exist for specific materials, those standards should be followed.

MEASUREMENT

Environmental Management Systems

Materials used in the project should originate from sources with a documented Environmental Management System (EMS). The EMS should cover the entire supply chain including acquiring and extracting raw materials, material manufacturing and associated processes, and related energy consumption. The EMS should adhere to the principles covered in ISO 14001.

Timber

All timber and wood products should originate from sustainably managed forests. All timber must be supplied by companies that hold Forest Stewardship Council (FSC) Chain of Custody Certification.

SUBMITTAL

For initial review, submit the Responsible Sourcing of Materials Calculator and the following supporting documents:

- Intended list of materials
- Documentation outlining the intended manufacturers and sourcing policies

For final review, submit the Responsible Sourcing of Materials Calculator and the following supporting documents:

- Bill of Quantities
- List of materials
- Specifications listing manufacturers
- Cost estimates
- Report providing sourcing policies of manufacturers
- Any documentation from manufacturers outlining third party sourcing standards such as FSC Chain of Custody Certification or ISO Certification

SCORE

Score	% (by cost) of Responsibly Sourced Materials (X)
-1	X < 20%
0	20% ≤ X < 30%
1	30% ≤ X < 40%
2	40% ≤ X < 50%
3	X ≥ 50%

MATERIALS [M.3] Recycled Materials

DESCRIPTION Encourage the use of materials made from recycled content in order to reduce the need for virgin materials.

MEASUREMENT PRINCIPLE All projects will use materials that are manufactured from recycled content.

MEASUREMENT All projects will complete the Recycled Materials Calculator to determine the percent, by cost, of the recycled content of all applicable materials based on the following measures: the percent of recycled content, the cost of materials with recycled content, and the total cost of applicable materials. Applicable materials include only materials used to develop infrastructure, public spaces, and shared facilities. All materials used to develop buildings or parks should be excluded and individually rated using GSAS Design, if available.

All projects will only consider materials permanently installed in the project. Exclude mechanical, electrical, and plumbing assemblies, as well as specialty items and equipment.

Materials which account for less than 10% by weight or volume of a component may be excluded.

SUBMITTAL For initial review, submit the Recycled Materials Calculator and the following supporting documents:

- Intended list of materials
- Documentation outlining all applicable materials and the percent of recycled content

For final review, submit the Recycled Materials Calculator and the following supporting documents:

- Bill of Quantities
- Cost estimates
- Manufacturer's documentation demonstrating the percent of recycled content

SUBMITTAL • Report outlining the use of recycled materials including the following information:
- Itemized list of recycled materials
- Description of the material
- Recycled content percentage of materials

Any materials claimed in [M.4] Materials Reuse may not be counted in this criterion.

SCORE

Score	% (by cost) of Recycled Materials (X)
-1	X < 5%
0	5% ≤ X < 10%
1	10% ≤ X < 15%
2	15% ≤ X < 20%
3	X ≥ 20%

MATERIALS [M.4] Materials Reuse

DESCRIPTION Encourage the reuse of elements and materials in order to reduce the need for virgin materials.

MEASUREMENT PRINCIPLE All projects will use building materials that are salvaged, reused, or refurbished from on- or off-site sources.

MEASUREMENT All projects will complete the Materials Reuse Calculator. The calculator computes the percent, by cost, of applicable materials reused based on the following measures: the quantity of reused materials, the reused material cost, and the total cost of applicable materials. Applicable materials include only materials used to develop public spaces and shared facilities. All materials used to develop buildings or parks should be excluded and individually rated using GSAS Design, if available. Reused materials can originate from on- or off-site sources .

All projects will only consider materials permanently installed in the project. Exclude mechanical, electrical, and plumbing assemblies, as well as specialty items and equipment.

Materials which account for less than 10% by weight or volume of a component may be excluded.

SUBMITTAL For initial review, submit the Materials Reuse Calculator and the following supporting documents:

- Intended list of materials
- Documentation outlining the intended reuse of materials from the project site or other sites

For final review, submit the Materials Reuse Calculator and the following supporting documents:

- Bill of Quantities
- Cost estimates
- Description of the material
- Description of how the material has been reused

SCORE

Score	% (by cost) of Materials Reuse (X)
-1	X < 5%
0	5% ≤ X < 10%
1	10% ≤ X < 15%
2	15% ≤ X < 20%
3	X ≥ 20%

DISTRICTS ASSESSMENT

MATERIALS [M.5] Life Cycle Assessment (LCA)

DESCRIPTION Encourage the use of materials and products which have the lowest life cycle environmental impact and embodied energy.

MEASUREMENT PRINCIPLE All projects will perform an eco-impact assessment based on the full life cycle of all materials from extraction to disposal.

MEASUREMENT Not scored for this phase as appropriate data to conduct a full Life Cycle Assessment is currently unavailable.

SUBMITTAL TBD

SCORE

Score	Requirement
-1	TBD
0	TBD
1	TBD
2	TBD
3	TBD

OUTDOOR ENVIRONMENT [OE]

The Outdoor Environment category consists of factors associated with outdoor environmental quality such as heat island effect, adverse winds, air flow, and acoustic quality within the district.

IMPACTS

Impacts resulting from ineffective control and design of the outdoor environment include:

- Climate Change
- Fossil Fuel Depletion
- Human Comfort & Health

MITIGATE IMPACT

Factors that could improve outdoor environmental quality include:

- Maximizing the amount of vegetation and solar reflectiveness to reduce the impact of a heat island effect
- Protecting spaces in the district from adverse wind conditions
- Ensuring a sufficient level of air flow to allow for the potential to naturally ventilate buildings
- Minimizing the amount of noise produced within the development

CATEGORY WEIGHT 7%

CRITERIA INCLUDED

No	Criteria	Min Score	Max Score
OE.1	Heat Island Effect	-1	3
OE.2	Adverse Wind Conditions	-1	3
OE.3	Air Flow	-1	3
OE.4	Acoustic Quality	-1	3
Total Possible		**-4**	**12**

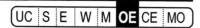
OUTDOOR ENVIRONMENT [OE.1] Heat Island Effect

DESCRIPTION Minimize heat island effect to reduce the impact on the surrounding habitat and environment.

MEASUREMENT PRINCIPLE All projects will develop strategies and perform calculations to ensure the heat island effect stays within a given threshold.

MEASUREMENT All projects will complete the Heat Island Effect Calculator to determine the additional heat island effect of the development. The project should specify the solar reflectance value and area for all surface areas within the site, including both ground surfaces and building rooftops. The project will calculate the overall area-weighted average solar reflectance of the site footprint for pre- and post-development. The difference between pre- and post-development overall solar reflectance values is used as an indicator to evaluate the performance of the development in minimizing the heat island effect.

SUBMITTAL Submit the Heat Island Effect Calculator and a site plan identifying the areas of ground and rooftop surfaces, and their corresponding reflectance values.

SCORE

Score	Performance Indicator (X)
-1	$X < -0.3$
0	$-0.3 \leq X < -0.2$
1	$-0.2 \leq X < -0.1$
2	$-0.1 \leq X < 0.0$
3	$X \geq 0.0$

OUTDOOR ENVIRONMENT [OE.2] Adverse Wind Conditions

DESCRIPTION   Minimize adverse wind conditions to surrounding spaces at the pedestrian level.

MEASUREMENT PRINCIPLE   All projects will develop strategies and perform wind control studies to minimize wind exposure to surrounding spaces at the pedestrian level.

MEASUREMENT   All projects will perform Computational Fluid Dynamics (CFD) simulations, pre- and post-development, to determine whether the increase in average wind speed, at the pedestrian level, is within the predefined reference threshold. Pre- and post-development average wind speeds should be located and measured at the corners of all collector and local road intersections. The reference threshold is defined as either an increase of 2 m/s or less from pre- to post-development OR a post-development wind speed of less than or equal to 5.5 m/s.

All projects will enter the results of the CFD simulations, for all locations, into the Adverse Wind Conditions Calculator. The calculator computes the final score by determining the number of locations falls within the predefined reference threshold.

SUBMITTAL   Submit the Adverse Wind Conditions Calculator and the following supporting documents:

- Plan drawings indicating measuring points
- Pre- and post-development simulation wind speed result plots
- Summary document of boundary conditions (Major wind speeds and directions)

SCORE

Score	% of Locations Demonstrating Compliance (X)
-1	X ≤ 50%
0	50% < X ≤ 60%
1	60% < X ≤ 70%
2	70% < X ≤ 80%
3	X > 80%

OUTDOOR ENVIRONMENT [OE.3] Air Flow

DESCRIPTION Promote air movement throughout the district to maximize the potential for natural ventilation within buildings.

MEASUREMENT PRINCIPLE All projects will develop a wind movement strategy to promote optimal air flow throughout the district.

MEASUREMENT All projects will use Computational Fluid Dynamics (CFD) simulations to determine whether the difference in average wind pressure throughout the district is within the predefined reference value. The average wind pressure should be measured at the midpoint between every intersection on all arterial roads.

All projects will enter the results of the CFD simulations, for all points, into the Air Flow Calculator. The calculator computes the final score by determining the percent deviation from the predefined reference value.

SUBMITTAL Submit the Air Flow Calculator and the following supporting documents:

- Plan drawings indicating measuring points
- Post-development pressure result plots
- Summary document of boundary conditions (Major wind speeds and direction)

SCORE

Score	% Deviation from Reference Value (X)
-1	X > 100%
0	75% < X ≤ 100%
1	50% < X ≤ 75%
2	25% < X ≤ 50%
3	X ≤ 25%

OUTDOOR ENVIRONMENT [OE.4] Acoustic Quality

DESCRIPTION Minimize the level of noise produced within the district affecting the development and surrounding environments.

MEASUREMENT PRINCIPLE All projects will develop a noise map and perform simulations to determine if any noise levels exceed the maximum zone noise limits of the district.

MEASUREMENT All projects will provide a strategic noise map of the development, showing the average day and night noise levels produced by all internal (within the district) and external (outside the district) sources. Internal sources should be derived from the development design, traffic plan, location of HVAC equipment on buildings, or other standard means supported by documentation. External sources can be determined from documents that describe existing or planned development. If the latter is unavailable, measurements performed for [UC.3] Acoustic Conditions can be used as input for the external sources in the calculation.

All projects will perform a simulation to determine the average day (7:00 - 22:00 hrs) and night (22:00 - 7:00 hrs) noise levels for each point of a dense grid for the development. The simulation should produce a noise map showing these levels, either as discrete point values or as continuous noise contours. All projects will include all relevant internal and external noise producing elements in the simulation, such as airports, highways, industrial zones, train stations, HVAC systems attached to buildings, district cooling plants, and street traffic within the development.

MEASUREMENT All projects will complete the Acoustic Quality Calculator to determine if any noise levels exceed the maximum noise limits by zone. Identify on the district zoning map the location of each of the four zone types: Industrial, Commercial, Residential and Quiet. Quiet zones include but are not limited to healthcare facilities. Educational institutions can be considered residential zones for the purposes of this criterion. Determine the highest occurring day and night level within each zone by overlaying the noise map on the zoning map. Noise levels are measured in dBA and averaged in Day (L_{day}) and Night (L_{night}) Levels.

If there are multiple areas of the same zone type, use the measurement from the zone with the highest average L_{day} and L_{night}. These noise levels should not exceed the following levels:

Zone Types	L_{day}	L_{night}
Industrial	70 dBA	65 dBA
Commercial	65 dBA	55 dBA
Residential	55 dBA	45 dBA
Quiet	45 dBA	35 dBA

Note: For districts with simple configurations, a series of manual calculations using acoustic handbooks may be sufficient. In more complex cases and especially when multiple line and point sources distributed within and outside the district are present, the project should conduct urban acoustic simulations.

DISTRICTS ASSESSMENT

SUBMITTAL For Initial Review, submit the Acoustic Quality Calculator and the following supporting documents:

- Strategic noise map, or the corresponding results of a noise simulation
- Plan identifying all noise producing elements, their location, and day and night intensities
- District map with zoning of the four types of acoustic zones; the zoning should be related to the master planning of different functions and spaces. For example, if a hospital facility is located in an otherwise commercial area, it should be separately identified, in this case as a "quiet" zone
- If a software is used to generate the sound map, the input values used in the simulation should be tabulated

For Final Review, verification must be provided in the form of measurements of existing noise levels. Standardized day and night measurements of noise levels should be performed on a randomly chosen normal operating day at selected locations. The outcomes for the equivalent L_{day} and L_{night} should be provided for all identified noise zones, for those locations where the noise map shows the highest occurring noise levels in that zone.

SCORE

Score	Requirement
-1	L_{day} and L_{night} exceed the allowed maximum noise levels in one or more zones.
3	L_{day} and L_{night} are within the allowed maximum noise levels for all zones.

CULTURAL & ECONOMIC VALUE [CE]

The Cultural and Economic Value category consists of factors associated with cultural conservation, support of the national economy, and diverse housing typologies.

IMPACTS

Impacts resulting from lack of cultural conservation, economic planning, and housing diversity include:

- Loss of Cultural Identity
- Economic Stagnancy or Decline
- Land Use and Contamination
- Long-term Viability of the District

MITIGATE IMPACT

Factors that could mitigate impact include:
- Encouraging designs to align with cultural identity and traditions
- Designing for a seamless integration into the existing cultural fabric
- Planning for the use of local materials and workforce
- Constructing a diverse mix of housing typologies

CATEGORY WEIGHT 13%

CRITERIA INCLUDED

No	Criteria	Min Score	Max Score
CE.1	Heritage and Cultural Identity	-1	3
CE.2	Support of National Economy	-1	3
CE.3	Housing Diversity	-1	3
Total Possible		**-3**	**9**

CULTURAL & ECONOMIC VALUE [CE.1] Heritage & Cultural Identity

DESCRIPTION Encourage design expressions that will align with and strengthen cultural identity and traditions.

MEASUREMENT PRINCIPLE All projects will develop a concept brief outlining design strategies. The Qatar Heritage Organization and/or an independent expert panel assigned by the Certification Authority will assess the design and determine whether the project meets the goals outlined in the mission statement.

MEASUREMENT All projects will develop a concept brief to demonstrate the following through drawings and descriptions of the design strategy:

- Enhancement, strengthening and reflection of cultural identity and traditions
- Harmonization with cultural values of the region

The compliance range will be defined by an Expert Heritage Panel assigned by the Certification Authority.

SUBMITTAL Submit a concept brief outlining design strategies that meet the criteria along with supporting design drawings or renderings to demonstrate design strategies.

SCORE

Score	Requirement
-1	TBD by Certification Authority
0	TBD by Certification Authority
1	TBD by Certification Authority
2	TBD by Certification Authority
3	TBD by Certification Authority

CULTURAL & ECONOMIC VALUE [CE.2] Support of National Economy

DESCRIPTION   Maximize the percentage of construction expenditures for goods and services originating from the national economy.

MEASUREMENT PRINCIPLE   All projects will maximize the percentage of construction expenditures benefitting the national economy.

MEASUREMENT   All projects will complete the Support of National Economy Calculator to indicate the amount of applicable construction expenditures benefitting the national economy, as a percentage of total construction costs. Applicable construction expenditures include only expenditures for infrastructure, public spaces, and shared facilities. All materials used to develop buildings should be excluded and rated individually with GSAS Design, if available. Construction expenditures include, but are not limited to, the following:

- Laborers/Contractors
- Infrastructure/Public Spaces/Shared Facilities Materials
- Construction Materials
- Construction Tools/Equipment
- Temporary Rental Spaces

SUBMITTAL   For initial review, submit the Support of National Economy Calculator, the intended list of materials, and the following supporting documents:

- Documentation outlining the intended contractual setup for construction expenditures
- Intended inventory of all materials
- Documentation outlining the intended estimate of construction expenditures with associated costs

DISTRICTS ASSESSMENT

UC | S | E | W | M | OE | **CE** | MO

SUBMITTAL For final review, submit the Support of National Economy Calculator and the following supporting documents:

- Report outlining the contractual setup for construction expenditures
- Inventory of all materials
- Comprehensive list of construction expenditures with associated costs

SCORE

Score	% of Construction Expenditure Benefitting National Economy (X)
-1	$X < 1\%$
0	$1\% \leq X < 10\%$
1	$10\% \leq X < 20\%$
2	$20\% \leq X < 30\%$
3	$X \geq 30\%$

CULTURAL & ECONOMIC VALUE [CE.3] Housing Diversity

DESCRIPTION Maximize the diversity of housing typologies within the district to ensure the long-term viability of the development.

MEASUREMENT PRINCIPLE All projects will develop a variety of housing typologies to promote a range of sustainable living options.

MEASUREMENT All projects will complete the Housing Diversity Calculator to determine the Simpson's Diversity Index and criterion score for the development. The calculator computes the score based on the quantity of different dwelling units for each of the following housing types:

- Single Family Detached
- Single Family Attached
- Multi-Family

SUBMITTAL Submit the Housing Diversity Calculator and supporting documentation outlining the number and type of each different dwelling unit.

SCORE

Score	Simpson Diversity Index (X)
-1	$X < 0.60$
0	$0.60 \leq X < 0.65$
1	$0.65 \leq X < 0.70$
2	$0.70 \leq X < 0.75$
3	$X \geq 0.75$

MANAGEMENT & OPERATIONS [MO]

The Management and Operations category consists of factors associated with design management and operations of the district.

IMPACTS

Environmental impacts resulting from ineffective building management and operations include:

- Climate Change
- Fossil Fuel Depletion
- Water Depletion
- Land Use & Contamination
- Water Pollution
- Air Pollution
- Human Comfort & Health

MITIGATE IMPACT

Factors that could mitigate environmental impact include:

- Creating a construction plan to mitigate negative effects of construction
- Creating a management plan to meet all the sustainable goals of the project
- Providing monitoring and management of wastewater facilities
- Providing facilities for the collection, storage, and proper removal of organic waste
- Providing facilities for the collection, storage, and proper removal of solid waste

CATEGORY WEIGHT 8%

CRITERIA INCLUDED

No	Criteria	Min Score	Max Score
MO.1	Construction Plan	0	3
MO.2	Management Plan	0	3
MO.3	Wastewater Management Plan	0	3
MO.4	Organic Waste Management Plan	0	3
MO.5	Solid Waste Management Plan	0	3
Total Possible		**0**	**15**

MANAGEMENT & OPERATIONS [MO.1] Construction Plan

DESCRIPTION
Encourage construction planning to minimize the adverse environmental effects caused by the construction process.

MEASUREMENT PRINCIPLE
All projects will develop and implement a Construction Plan for the responsible and sustainable management of the construction process.

MEASUREMENT
The Construction Plan will include provisions and indicate compliance for the following categories:

- Noise Pollution
- Dust Control
- Water Contamination
- Waste Management
- Recycling Management

For each of the above items, the Construction Plan should identify potential sources of environmental impacts, as well as strategies for how those impacts will be mitigated. The plan can reference the GSAS Construction Assessment and Guidelines or identify other standards and references to which construction activity will adhere.

SUBMITTAL
Submit the following supporting documents as part of a comprehensive Construction Plan:

- Project requirements and design intent
- Documents demonstrating compliance for all essential elements of the Construction Plan

SCORE

Score	Requirement
0	Construction Plan does not demonstrate compliance.
3	Construction Plan demonstrates compliance.

MANAGEMENT & OPERATIONS [MO.2] Management Plan

DESCRIPTION   Encourage management planning of the development during the design phase to ensure continued sustainability during the operation of the district.

MEASUREMENT PRINCIPLE   All projects will develop and implement a Management Plan for all phases of the district's operation.

MEASUREMENT   The Management Plan will demonstrate the intent to monitor and manage community-wide systems as well as ensure long term sustainability through implementing maintenance plans and awareness programs. The plan should indicate compliance in the following categories:

- Water Management Systems
- Energy Management Systems
- Transportation and Infrastructure Systems
- Landscape Maintenance
- Information Systems
- Sustainable Awareness and Education

SUBMITTAL   Submit the following supporting documents as part of a comprehensive Management Plan:

- Project requirements and design intent
- Documents demonstrating compliance for all essential elements of the Management Plan
- Documents that outline steps necessary for continued management during the operation of the district

SCORE

Score	Requirement
0	Management Plan does not demonstrate compliance.
3	Management Plan demonstrates compliance.

MANAGEMENT & OPERATIONS   [MO.3] Wastewater Management Plan

DESCRIPTION   Encourage infrastructure planning to designate containment and treatment facilities for the district's wastewater streams in order to minimize water pollution and provide recycled water for reuse.

MEASUREMENT PRINCIPLE   All projects will develop and implement a Wastewater Management Plan for the collection, storage, treatment, and reuse of wastewater.

MEASUREMENT   The Wastewater Management Plan will include provisions for managing wastewater produced by occupants, as well as for managing recycled water used for occupants and irrigation. The plan should indicate compliance in the following categories:

- Blackwater sewage system and treatment facility
- Greywater sewage system and treatment facility
- Recycled water strategies for reuse
- Leak detection

SUBMITTAL   Submit the following supporting documents as part of a comprehensive Wastewater Management Plan:

- Project requirements and design intent
- Documents demonstrating compliance for all essential elements of the Wastewater Management Plan
- Documents that outline steps necessary for continued wastewater management during the operation of the district

SCORE

Score	Requirement
0	Wastewater Management Plan does not demonstrate compliance.
3	Wastewater Management Plan demonstrates compliance.

MANAGEMENT & OPERATIONS [MO.4] Organic Waste Management Plan

DESCRIPTION Encourage planning for the collection and composting of organic waste in order to ensure continued sustainability of the district.

MEASUREMENT PRINCIPLE All projects will develop and implement an Organic Waste Management plan for the collection, removal, and composting of organic waste.

MEASUREMENT The Organic Waste Management Plan will include provisions for managing organic waste produced by users. The plan should indicate compliance in the following categories:

- Infrastructure for collecting, removing, and composting organic waste
- A composting facility and its intended capacity which can meet the needs of the development
- Strategies and policies for reusing or distributing organic waste

SUBMITTAL Submit the following supporting documents as part of a comprehensive Organic Waste Management Plan:

- Project requirements and design intent
- Documents demonstrating compliance for all essential elements of the Organic Waste Management Plan
- Documents that outline steps necessary for continued organic waste management during the operation of the district

SCORE

Score	Requirement
0	Organic Waste Management Plan does not demonstrate compliance.
3	Organic Waste Management Plan demonstrates compliance.

MANAGEMENT & OPERATIONS [MO.5] Solid Waste Management Plan

DESCRIPTION Encourage recycling and waste collection planning in order to minimize waste taken to landfills or incineration facilities.

MEASUREMENT PRINCIPLE All projects will develop and implement a Solid Waste Management Plan for the collection, storage, and removal of waste.

MEASUREMENT The Solid Waste Management Plan will include provisions for managing solid waste. The plan should indicate compliance in the following categories:

- Infrastructure plans for collecting, storing, and treating solid waste on-site
- Plan for solid waste leaving the site, including:
  - Locations and intended capacities of storage facilities
  - Type and locations of disposal facilities
- Recycling Plan, if applicable, including strategies for the collection, storage, and reuse of recycled materials

SUBMITTAL Submit the following supporting documents as part of a comprehensive Solid Waste Management Plan:

- Project requirements and design intent
- Documents demonstrating compliance for all essential elements of the Solid Waste Management Plan
- Documents that outline steps necessary for continued solid waste management during the operation of the district

SCORE

Score	Requirement
0	Solid Waste Management Plan does not demonstrate compliance.
3	Solid Waste Management Plan demonstrates compliance.

Made in the USA
Charleston, SC
01 August 2013